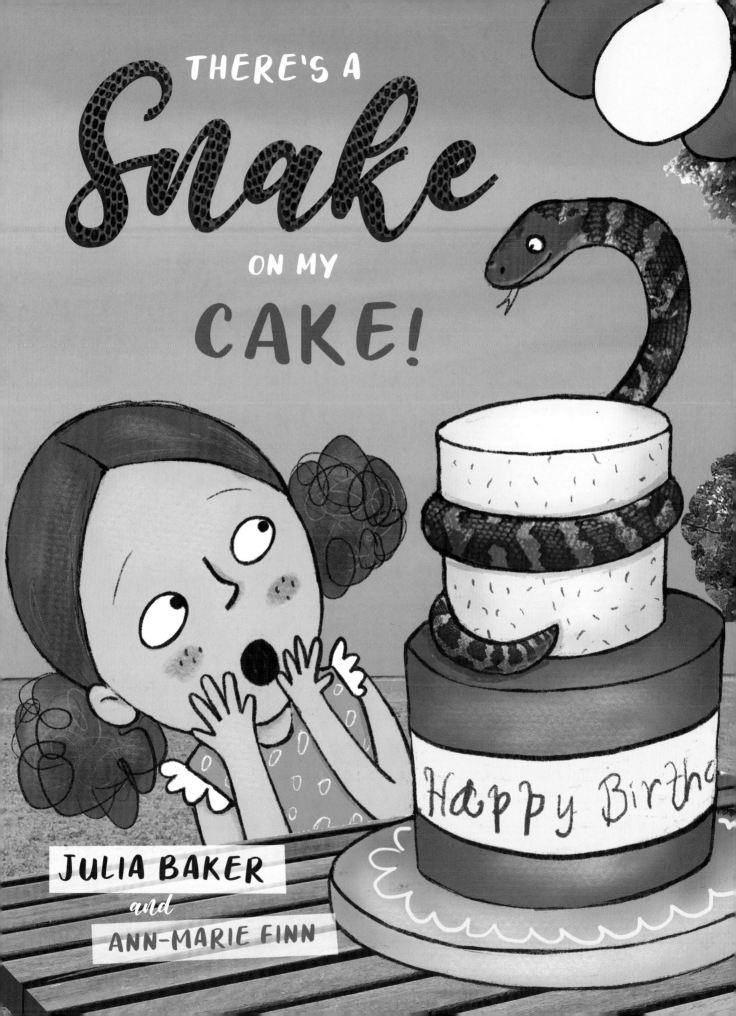

There's a Snake on my Cake
Text © by Julia Baker 2020
Illustrations © by Ann-Marie Finn 2020
Published by Wombat Books, 2020
P.O. Box 1519, Capalaba, QLD 4157
www.wombatbooks.com.au
info@wombatbooks.com.au

978-1-925563-94-8
A Cataloguing Record for this book is available at the
National Library of Australia.

THERE'S A
Snake
ON MY
CAKE!

JULIA BAKER
and
ANN-MARIE FINN

It was almost Jessie's birthday
and she was planning a party
with all her friends.

Annabelle had
a unicorn party.

Everyone
pranced
and
danced
with sparkly horns,

before
they sat
down to
fairy bread.

Nathan went rollerblading for his birthday.

Everyone rolled and wobbled around the rink.

Before they sat down
to sausage
rolls.

Phoebe's sixth birthday was
a football game.

Everyone kicked and
flicked the ball around,

Before they sat down
to orange wedges and cake.

Jessie had enjoyed prancing and dancing,

Rolling and wobbling,

Even kicking and flicking...

But she wanted something
DIFFERENT
for her birthday.

Jessie wanted a...

SNAKE PARTY!

Jessie wondered how to show her friends that a snake birthday party was just as cool as unicorns or rollerblading or football.

Then it came to her...

She needed expert help.

On the day of
the party, Jessie waited
nervously at the door.

Her friends arrived,
excited to see who
the surprise guest might be.

It was Julia, the snake catcher, with some guests of her own!

They met a black and gold jungle python...

A spotted python...

And the star of the show...

Mango!

An albino carpet python!

Everyone ooohed and aaahed
at the slithering spectacle,

Before they played
pin-the-tail
on the
rattlesnake

'And now for the BEST snake of all,' announced Julia.